CRYPTID GUIDES: CREATURES OF FOLKLORE

GUIDE TO CHUPACABRAS

A Crabtree Branches Book

BY CARRIE GLEASON

Crabtree Publishing
crabtreebooks.com

Developed and produced by Plan B Book Packagers
www.planbbookpackagers.com
Art director: Rosie Gowsell Pattison

Crabtree editor: Ellen Rodger
Prepress technician: Margaret Salter
Production coordinator: Katherine Berti
Proofreader: Melissa Boyce

Photographs:
Pg 8 Rosie Gowsell; all other images Shutterstock.com.

Crabtree Publishing

crabtreebooks.com   800-387-7650
Copyright © 2023 Crabtree Publishing

Hardcover	978-1-0396-6343-5
Paperback	978-1-0396-6392-3
Ebook (pdf)	978-1-0398-0718-1
Epub	978-1-0398-0745-7

Published in Canada
Crabtree Publishing
616 Welland Avenue
St. Catharines, Ontario
L2M 5V6

Published in the United States
Crabtree Publishing
347 Fifth Avenue
Suite 1402-145
New York, NY 10016

Library and Archives Canada
Cataloguing in Publication
Available at Library and Archives Canada

Library of Congress
Cataloging-in-Publication Data
Available at the Library of Congress

Printed in the U.S.A./012023/CG20220815

CONTENTS

CRYPTIDS and KINFOLK

This chart shows some of the best-known cryptids and creatures from folklore. How many of them do you think are real?

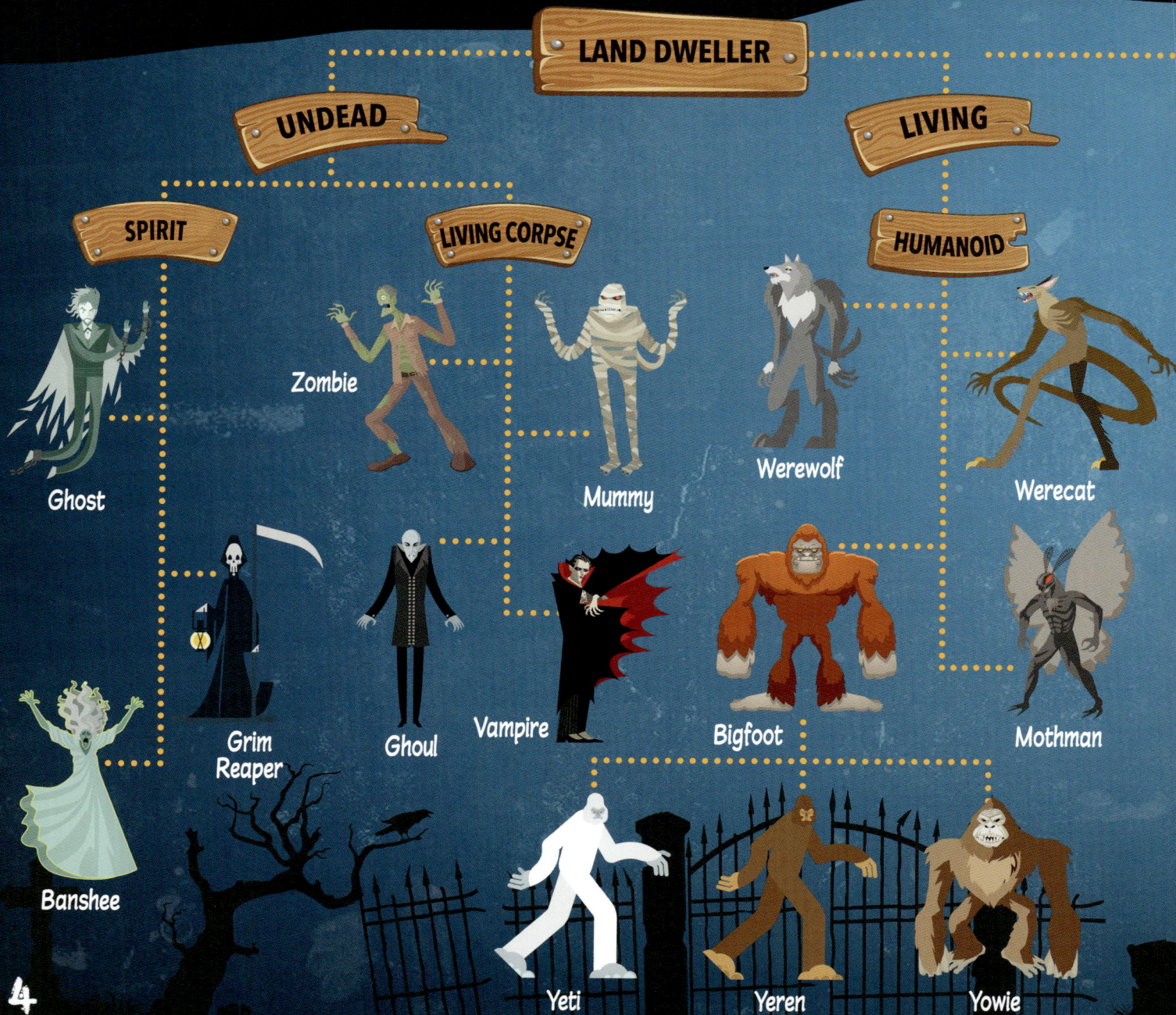

LAND DWELLER

UNDEAD

LIVING

SPIRIT

LIVING CORPSE

HUMANOID

Ghost

Zombie

Mummy

Werewolf

Werecat

Banshee

Grim Reaper

Ghoul

Vampire

Bigfoot

Mothman

Yeti

Yeren

Yowie

WHAT IS A CHUPACABRA?

A chupacabra is a cryptid—a creature from folklore whose existence is not yet proven to be true. Cryptozoologists are people who search for and study these creatures. They gather stories from folklore and investigate reported sightings of cryptids. This book shows what is true, what is thought to be true, and what are outright lies and hoaxes about chupacabras. Could chupacabras be real? This Cryptid Guide will help you decide!

ANIMAL & HYBRID

Unicorn

Jackalope

Chupacabra

OTHERWORLDLY

GIANT

Fairy

Elf

Nymph

Ogre

Troll

EXTRATERRESTRIAL

SEA MONSTER

Alien

SWAMP MONSTER

Mermaid

Sea Serpent

LAKE MONSTER

Louisiana Swamp Monster

Leviathan

Loch Ness Monster

Ogopogo

Champ

5

FEAR THE CHUPACABRA..

Imagine this: It's not too long ago, and you live on the Caribbean island of Puerto Rico. In some places around the world, people have just started using the Internet and cell phones, but where you live, people stick to the old ways. You're standing on the side of the road with a crowd of townspeople. All around, there is talk of mysterious things that have been happening. Livestock have been found dead, drained of their blood. A cheer rises from the crowd as pickup trucks appear on the road. In the back of one of them is a cage with a goat inside. The goat will be used as live bait to try to catch the creature that killed the livestock. The goat lets out a frightened bleat, and you wonder if it will make it back…alive.

CHUPACABRA BASICS

Reports of a bloodsucking cryptid called a chupacabra first surfaced in Puerto Rico in 1995 after livestock were found drained of their blood and an eyewitness reported seeing an alien-like creature outside her house. News of the chupacabra spread around the world. Since then, discoveries of dead livestock have led to claims of chupacabra attacks in Mexico, Texas, Florida, and countries in Central and South America. The biggest mystery about chupacabras is what they look like. Eyewitnesses describe them in different ways, from an alien-like reptile to a dog-like creature. But one thing is for certain—rumors of chupacabras strike fear into the hearts of farmers and small pet owners.

TOP CRYPTIDS

Bigfoot

Bigfoot are North America's most famous cryptids. While they are most commonly believed to live in the forests of the Pacific Northwest, sightings have been reported across the continent.

Loch Ness Monster

The Loch Ness Monster, or Nessie, is a lake monster that is believed to live in Scotland. Nessie is the most hunted cryptid in the world. Thousands of people travel to Loch Ness (a lake in Scotland) each year to try and get a glimpse of it.

Aliens

Some cryptids are believed to be creatures from outer space. This could explain why some cryptids seem to have special abilities and can stay so well hidden. The Dover Demon is an alien cryptid that was sighted for two days in Massachusetts in 1977, before disappearing. It has not been spotted since.

ANATOMY OF A CHUPACABRA

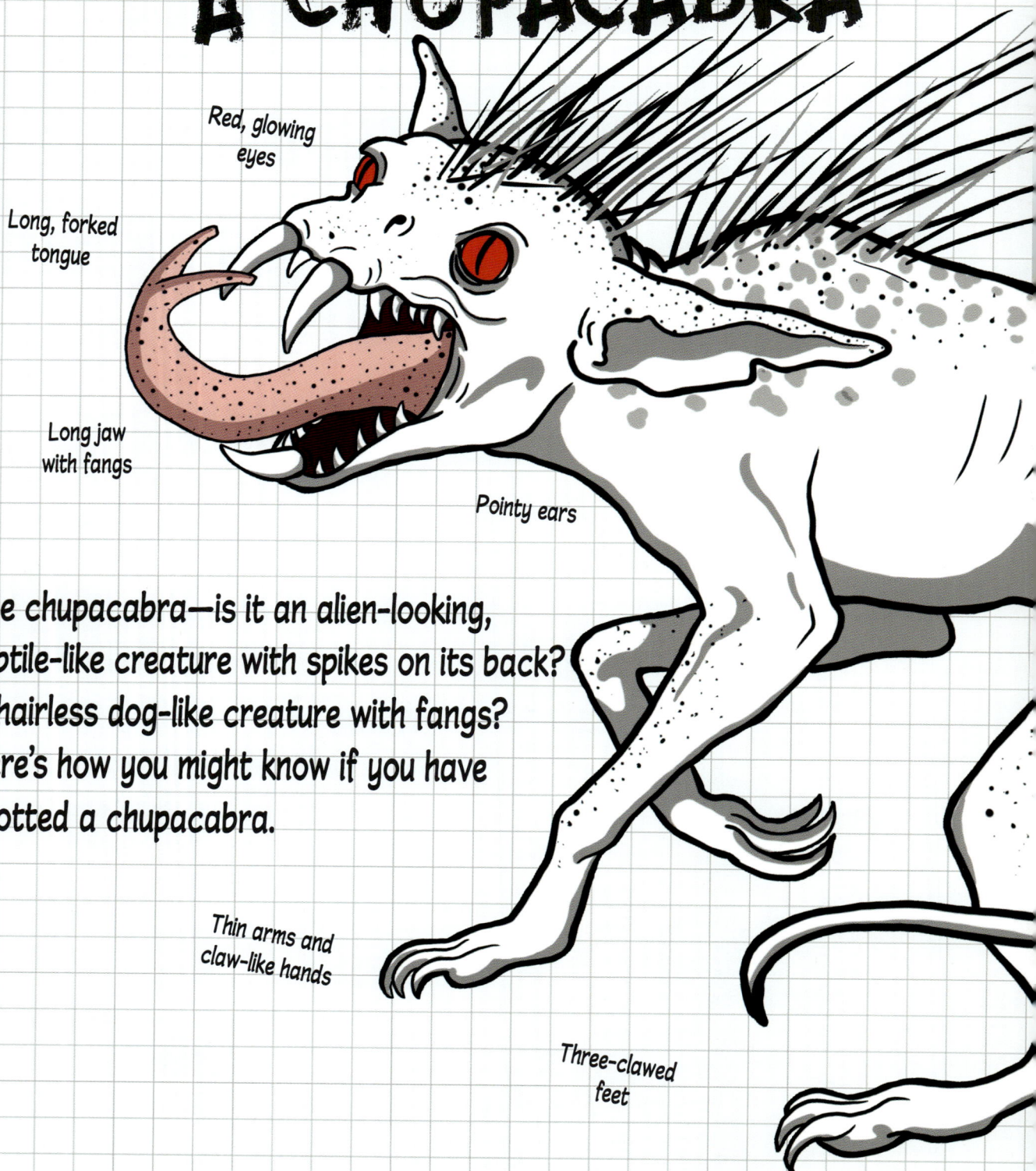

Red, glowing eyes

Long, forked tongue

Long jaw with fangs

Pointy ears

The chupacabra—is it an alien-looking, reptile-like creature with spikes on its back? A hairless dog-like creature with fangs? Here's how you might know if you have spotted a chupacabra.

Thin arms and claw-like hands

Three-clawed feet

Leathery black, green, gray, or brown skin

4 to 5 feet (1.2 to 1.5 m) tall

Oval-shaped head that may look like a reptile's, a dog's, or a panther's head

Spines or ridges down its neck and spine

May have large, bat-like wings

Powerful goat- or kangaroo-like hind legs

THE FRANKENSTEIN OF CRYPTIDS?

Descriptions of the chupacabra seem to "borrow" body parts from many different creatures. In folklore, a chimera is an imaginary monster made up of parts from different animals. If the chupacabra is a chimera, it would be made up of different parts of some—or all!—of these creatures.

 KOMODO DRAGON

+

 ALIEN

+

 KANGAROO

+

 BAT

+

 DOG

+

 VAMPIRE

THE WAY OF THE CHUPACABRA

The telltale sign that a chupacabra is nearby is the dead livestock it leaves behind. No humans have been attacked by a chupacabra, so you're safe—for now! Cryptozoologists have been able to piece together the following information about the mysterious chupacabra from different suspected chupacabra attacks.

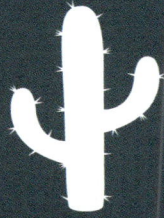

HABITAT

Chupacabras live in rain forests, deserts, grasslands, and plains.

DIET

Chupacabras are said to drink the blood of livestock such as goats, sheep, chickens, pigs, and cattle, and pets such as cats and small dogs. They often leave puncture marks on their prey, and in some cases have ripped open their throats.

BEHAVIOR

Chupacabras are said to…

- be nocturnal, which means they are active at night
- fly using wings, or hop or jump long distances using their powerful hind legs
- move incredibly fast
- make hissing and screeching sounds
- smell like rotten eggs
- be solitary creatures, which means they live and hunt alone

WHAT TO DO IF YOU SEE A CHUPACABRA

Eyewitnesses say that chupacabras are easily frightened away by humans, so if you see one, make yourself bigger by standing up tall and raising your arms. You can also shout, holler, stomp your feet, and clap your hands, as these loud noises may frighten it off.

OTHER BLOODSUCKING CRYPTIDS

Chupacabras are said to drink the blood of their prey using their sharp fangs. Folklore around the world is filled with stories of bloodsucking monsters. But unlike the chupacabra, these ones have a taste for human blood.

VAMPIRES

Vampires come from the folklore of Eastern Europe. Human-like in appearance, they are said to have risen from the dead. The fear of vampires was so real in Europe in the 1700s that corpses were dug up so that stakes could be driven through their hearts or their heads could be cut off to keep them from terrorizing villages.

BAOBHAN SITH

A baobhan sith is a bloodsucking fairy-vampire from Scottish legends that always wears a green dress. They have deer hooves for feet and are nocturnal. They can change into a raven, a crow, or a wolf. Instead of fangs, they use their fingernails to scratch their victims to get at their blood.

YARA-MA-YHA-WHO

A yara-ma-yha-who is a vampire-like creature from Australian legends. It has red skin and is described as a cross between a frog and a human. It has a large mouth but no teeth. Instead, it uses suckers on its fingers and toes to drain blood from its victims. The creature perches in tree branches and drops down on victims from above.

LOOGAROO

A loogaroo is a bloodsucking hag from Caribbean folklore. These shapeshifting **streams of blue light** take the forms of old women during the day by wearing their skin. People become loogaroos by making a deal with the devil, who gives them magic powers in return for providing him with blood. Instead of sucking on necks like vampires, they suck blood from a sleeping person's arms or legs.

THE LEGEND BEGINS

Strange things were happening to animals in Puerto Rico in the early 1990s. Something was killing farmers' livestock. Then, in 1995, 150 livestock were found dead around the town of Canóvanas in a single year. Who—or what—was to blame?

A SCARY SIGHT

One day in late summer 1995, a woman named Madelyne Tolentino was woken up from a nap by her mother. She had seen a strange creature outside their house. When Madelyne went to look, she saw something she had never seen before. It was a creature about 3 feet (0.9 m) tall, standing on two legs like a human. It had gray, leathery skin and spikes on its back. The creature had large eyes, and no nose except for dots for nostrils. Madelyne watched the creature for a few minutes before it hopped away. She told newspapers what she had seen. Before long, the creature was given the name "chupacabra," which means "goat sucker" in Spanish.

THE HUNT IS ON

Soon, other people came forward saying that they too had seen the creature. But each time, details about the chupacabra's appearance changed a bit. Some people said it could fly through the air, and had a long, forked tongue like a snake. It could also change color like a chameleon. Authorities in Puerto Rico took the reports of the creature seriously. The mayor of Canóvanas set up search parties to find and capture it. Although they searched for a year, the chupacabra that terrorized Puerto Rico in 1995 was never found.

Movie monster or real monster?

The Puerto Rican chupacabra is more alien-like in appearance than chupacabras sighted in other countries. Some cryptozoologists think that Madelyne Tolentino's description matches the alien from the movie *Species*, which had just been released that year.

Twenty years before the first chupacabra report, Puerto Rico was plagued by another vampire beast called the Moca Vampire. The Moca Vampire was said to be a large, feathered creature that could fly. Around the same time, there were many UFO sightings around Puerto Rico. Could the Moca Vampire have been a feathered, flying alien?

CHUPACABRAS AROUND THE WORLD

News of the search for Puerto Rico's chupacabra spread quickly around the world. In the years that followed, sightings were reported as far away as Russia and the Philippines. Some people wondered how the new cryptid had spread around the world so quickly. Others wondered if the chupacabra was being blamed for any mysterious animal deaths that occurred. This map shows the year that chupacabras were first reported in different countries.

North America

Texas, 1996

Florida, 1996

Mexico, 1996

Puerto Rico, 1995

South America

Brazil, 1999

Chile, 2000

Argentina, 2002

Russia, 2011

Asia

urope

Africa

India, 2018

Sri Lanka, 2016

Philippines, 2016

Australia

OUT-OF-THIS-WORLD THEORIES

Over the years, people have come up with some far-out theories, or ideas, to try and explain chupacabras. They range from aliens to top-secret military experiments. These kinds of explanations usually involve governments keeping secrets from the public, and are called conspiracy theories.

Theory: Chupacabras are a top-secret military weapon

Explanation: People who believe this theory say that the U.S. military is secretly creating "super-soldiers" at a military base in Puerto Rico's El Yunque rain forest, and that the chupacabra is one of these secret super-soldiers that has escaped.

Date: 1995

Date: 1974

Theory: Chupacabras are aliens

Explanation: A powerful message was beamed into space from the Arecibo Observatory in Puerto Rico by the Search for Extraterrestrial Intelligence (SETI) Institute. Some people claim that communication with aliens was secretly successful and that since then, UFOs often visit Puerto Rico. Believers of this theory say that chupacabras are aliens that UFOs have left behind and that they have been in Puerto Rico since the 1970s.

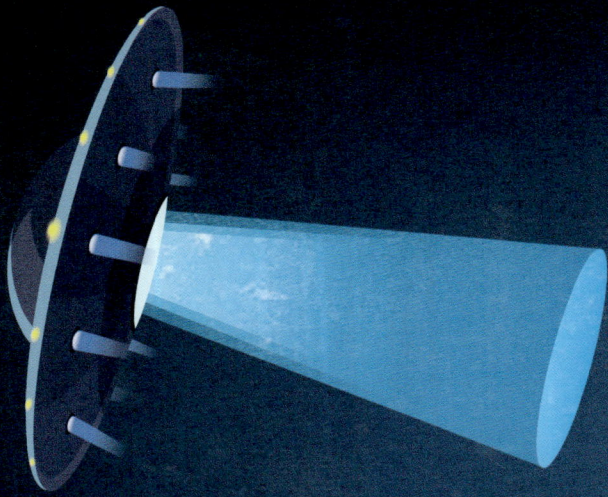

Date: 2000

Theory: **Governments are hiding proof of chupacabras**

Explanation: In Nicaragua, a farmer named Jorge Luis Talavera shot and killed a chupacabra that he claimed killed 25 of his sheep. The animal carcass was sent to a university medical college, where tests showed it was a dog. But when the chupacabra's body was returned to Talavera, he said it wasn't the same one he had sent for testing. He claimed the college had kept the real chupacabra and sent it to the government as part of a cover-up.

According to some UFO researchers from Chile, chupacabras hatch from eggs.

Date: 2000

Theory: **Chupacabras are a NASA experiment**

Explanation: According to this theory, soldiers discovered a family of chupacabras living in a mine in northern Chile. Scientists from NASA immediately took the chupacabras. Believers say that the chupacabras had escaped from a secret lab in the Atacama Desert, where NASA scientists were breeding them to create creatures that could survive on Mars.

CANINE CHUPACABRAS

When stories of chupacabra sightings began in the southern United States and Mexico in 1996, the creature was described as having a more dog-like appearance than an alien appearance. This led cryptozoologists to wonder: Did the chupacabra change its looks, or is it a completely different cryptid?

MEET THE CANINES

The canine family includes 34 dog-like animals that are closely related. Canines include domestic dogs (like your pet dog), wild dogs, wolves, coyotes, jackals, and foxes. Sometimes different breeds of dogs are mixed together on purpose to create new dog breeds, such as a labradoodle, which is a mix of a lab and a poodle. In nature, different canine species can also sometimes mix. For example, a coyote and wolf mix is called a coywolf. Some people think that the chupacabra found in Mexico and the United States may be a new canine mix.

DOG DNA PROOF

In Texas, several suspected chupacabras have been caught, dead and alive, by farmers and ranchers after their livestock have been attacked. DNA testing has shown that these chupacabras are actually dogs, coyotes, or some other canine mix.

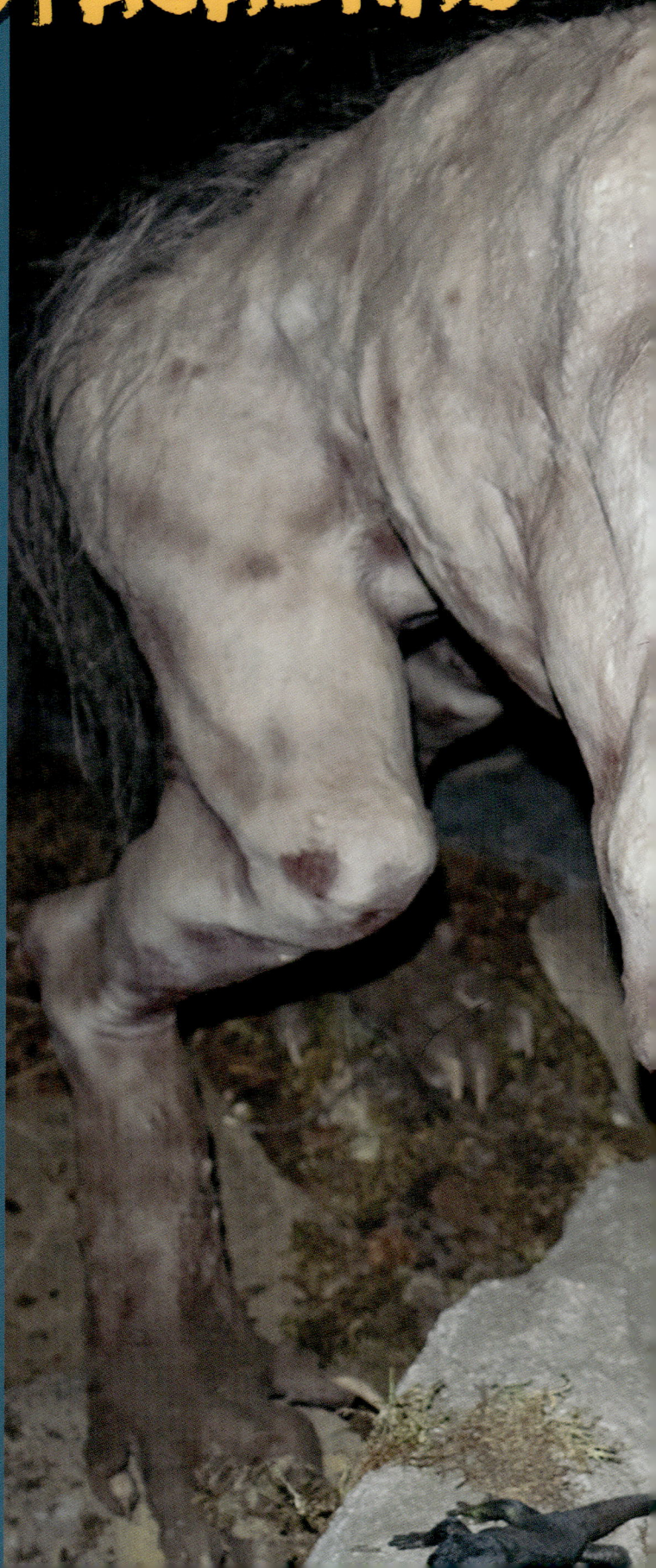

TEXAS CHUPACABRA

Chupacabra sightings are common in Texas, where they are also sometimes referred to as Blue Dogs. Some cryptozoologists think that Blue Dogs are wild Xoloitzcuintli, or Mexican hairless dogs. These dogs have no fur on their bodies except for a small tuft on their heads. They have black or bluish-gray skin. The Xoloitzcuintli dog breed was considered sacred in the ancient Aztec culture of Mexico. The breed was almost wiped out by Spanish settlers, who ate them. Since the 1960s, their population has recovered and these dogs now roam close to the U.S.–Mexico border.

REAL DOG

HELLHOUNDS

Dogs may be humans' best friends, but in some myths and legends, dogs and other canine creatures are anything but friendly. In folklore, hellhounds are dog-like creatures that come from the underworld. They are usually depicted as large black dogs with glowing red eyes. Cultures around the world have legends relating dogs to death.

CERBERUS

In Greek mythology, a dog named Cerberus guards the entrance to the underworld, the place where the dead go. Its job is to keep the dead from leaving. Cerberus is described as having three heads and the tail of a snake.

GRIMS

English legends are filled with stories of ghostly black dogs. These creatures are often an omen of death, but can also appear as protectors and guides for travelers and lost children. A special type of this black dog is the Church Grim, a dog buried in a Christian church's graveyard. The dog's ghost, called the Grim, protects the graves from vandals and evil spirits.

CADEJOS

In Central American folklore, there are two dog-like supernatural beings called cadejos. These large shaggy dogs have red eyes and hooves like a goat or deer. One cadejo is black and the other is white. They appear to travelers at night. Seeing the white cadejo brings good luck, but seeing the black cadejo is a warning of death.

XOLOTL

According to legends of the Aztec people who lived in what is now Mexico, Xolotl was the god of fire and lightning. Xolotl was shown in art as having a human body and a dog-like head. Xolotl guided the spirits of the dead to Mictlan, the Aztec underworld. The Xoloitzcuintli dog breed is named after this god.

CHUPACABRAS MAKE IT BIG

In popular culture, the mystery around what type of creature the chupacabra might be includes not only dogs, aliens, and vampires, but also werewolves and even a type of bigfoot! This timeline shows some of the chupacabra's biggest hits in recent movies, TV shows, and books.

2013
One of the main characters in *Planes*, an animated movie about talking airplanes, is named El Chupacabra. He says his name is meant to "strike fear in the hearts of his opponents."

2013
Chupacabra, the third book in the children's series Cryptid Hunters by Roland Smith, is published.

2014
In the horror movie *Indigenous*, a chupacabra haunts a tropical forest in Panama, a country in Central America. Tourists ignore the warnings from locals not to visit the forest.

2017
A children's picture book called *The Chupacabra Ate the Candelabra* is published. In the story, three goats go looking for a chupacabra before it can find and eat them, only to learn that the chupacabra prefers goat cheese to goat's blood.

START HERE

1995
Early descriptions of the chupacabra are very similar to the alien in the sci-fi movie *Species*, which was released the same year as the first chupacabra sighting.

1997
The popular TV show *The X-Files* makes an episode called "El Mundo Gira," in which a missing Mexican worker is thought to have been the victim of a chupacabra attack.

2001
In the *Looney Tunes* webtoon episode "El Chupacabra," Daffy Duck and Porky Pig are cryptid hunters investigating the chupacabra.

2005
In the horror movie *Mexican Werewolf in Texas*, the chupacabra is a werewolf.

2003
In *Scooby-Doo! and the Monster of Mexico*, the chupacabra is described as a Mexican bigfoot. It has an ape-like appearance.

2018
The picture book *El Chupacabras* is published. It is about a girl who lives on a goat farm. The book is written in English and Spanish. In the story, the chupacabra saves the town after the goats are made into giants by a magic powder meant to protect them.

2021
A short children's film called *The Last of the Chupacabras* is released. It is about a chupacabra marionette that comes to life during an earthquake and keeps a lonely old woman company.

CHUPACABRAS VS DOGMEN

Be careful not to confuse the dog-like chupacabra with another type of cryptid commonly sighted in North America—dogmen. Dogmen are humanoid **cryptids**, which means they have human-like characteristics, while chupacabras do not. Here's how chupacabras and dogmen compare.

CHUPACABRA

1 Chupacabras are often said to be hairless or have hairy patches on their skin. Dogmen are covered in hair from head to toe.

DANGER: HUMANOID CRYPTIDS

GOATMEN

DANGER METER

Goatmen are half human and half goat. The most famous North American goatman is said to live in Maryland, and is more of a danger to dogs than to people, although it has been said to terrorize teenagers kissing in cars.

GATORMEN

DANGER METER

A gatorman is half alligator and half human. Gatormen have been reported in the southeastern United States, where they live in swamps. Although people do sometimes go missing in swamps, gatormen have not yet been blamed for their disappearances.

DOGMAN

2 Dogmen may have human-looking eyes and a human-sounding voice.

3 Dogmen may have paws that are more similar to human hands than animal paws.

4 Both dogmen and chupacabras have been reported to walk or stand on two legs, but dogmen are said to walk like humans.

5 Dogmen are said to be as tall as 7 feet (2 m), which is taller than most reported chupacabras.

MOTHMEN

DANGER METER

Mothmen are half human and half winged animal. They are believed to be a sign that a disaster is coming. The most famous mothman is from West Virginia.

APEMEN

DANGER METER

Apemen are half ape and half human. Bigfoot are the most famous of the apemen. There have been reports of suspected bigfoot attacks on people.

DEER PEOPLE

DANGER METER

Deer people either have the horns and face of a deer with the body of a human, or a human face and the legs and hooves of a deer. They have been sighted at roadsides all over North America and, like real deer, may be responsible for numerous traffic accidents.

THE CRYPTID RECORD

Cryptozoology's #1 Source for Sightings

Dogmen Sightings Across America

Reported Sightings

Across the country, people have claimed to have spotted dogmen. Reports of sightings are important for cryptozoologists. They interview people who have encountered cryptids and read about past sightings to try to put together the full story of a cryptid. These are some descriptions of dogmen from real-life sightings.

Grunch Road Monster
New Orleans

Increased sightings of the "Grunch" were reported all over New Orleans following Hurricane Katrina in 2005. It is believed that the creatures, which resemble a mix between a reptile-like chupacabra and a dog-like chupacabra, were forced out of their homes in the swamps due to flooding. They are about 3 feet (0.9 m) tall with scaly skin and patchy hair, horned ears, and red or green glowing eyes.

Beast of Bray Road
Wisconsin

First sighted in 1936 in Elkhorn, Wisconsin, the Beast of Bray Road is about 6 feet (1.8 m) tall, has gray or brown fur, a wolf-like face, the body of a human, and the tail of a German shepherd. Most sightings have occurred at roadsides, after which the creature disappears into the woods, but it has also been said to have damaged cars with its long, sharp claws.

Michigan Dogman
Michigan

There have been numerous sightings of the Michigan Dogman. Witnesses say that it is 7 feet (2 m) tall, walks on two legs, and has the upper body of a man. Its howl sounds like a human scream. According to legend, the Michigan Dogman appears every 10 years and is due to appear again in 2027.

(left) A 3-D computer drawing of the Michigan Dogman in the wilderness of northern Michigan. It was first sighted in 1887.

Minnesota Dogman
Minnesota

In 2009, a man claimed he spotted the Minnesota Dogman in a forest while it was hunting deer. The eyewitness said that the dogman stood on two legs, which were large and muscular like a dog's. The creature had a dog-like face and hands like a human, and it left paw prints in the snow.

Chupacabras Explained

Several animal carcasses thought to be the dead bodies of chupacabras have been examined and found to be dogs or coyotes with a skin disease called mange. This disease causes patchy hair loss, a thickening of the skin that makes it look like leather, and general weakness. Weakness means that the animal can't hunt well, making it more likely to prey on livestock.

What Killed the Livestock?

It is possible that many deaths blamed on chupacabras may have been from wild dogs. Dogs have an **instinct** that tells them to chase, attack, and kill other animals, even if they don't eat them. Dogs are also known to attack the necks of their victims. Some people also think that the livestock deaths in Puerto Rico may have been caused by weasel-like animals called mongooses. These are known to be vicious and attack animals larger than themselves.

What About the Blood Loss?

When an animal dies, the heart stops pumping blood around the body. Blood collects at the lowest points of the body, causing skin in other parts to change color, and look as though the animal has lost blood. This stage of death is called livor mortis. The livestock believed to be victims of chupacabra attacks may have been in this stage of death when they were discovered.

Vampire Animals

In nature, some real animals do drink the blood of other animals. These blood-drinking animals are known as sanguivorous animals. Here are just a few of them.

Leeches

Vampire bats

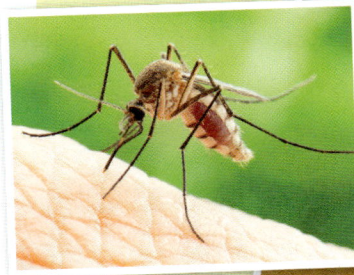
Mosquitoes

Bedbugs

Lampreys

LEARNING MORE

Want to know more about cryptids, myths, and monsters such as the ones described in this book? Here are some resources to check out on your cryptid-hunting quest.

Books

Encountering Chupacabra and Other Cryptids: Eyewitness Accounts by Megan C. Peterson. Capstone Press, 2015.

Searching for El Chupacabra by Jennifer Rivkin. PowerKids Press, 2015.

TV and Films

Monstrum is a series of videos created by PBS about monsters, myths, and legends.

Find the videos on the PBS website at:

www.pbs.org/show/monstrum/

Websites

The Centre for Fortean Zoology is a cryptozoology organization that researches cryptids from around the world. They produce a weekly TV show, books, and magazines about cryptids.

www.cfz.org.uk/

GLOSSARY

Aztec An Indigenous people who spoke the Nahuatl language and established a large empire in Mexico that lasted from 1345 to 1521

breed Types of animals that all share general characteristics, such as appearance

carcass The body of a dead animal

corpse A dead body

DNA Molecules, or small particles, in cells that contain genetic code

folklore The stories, customs, and beliefs that people of a certain place share and pass down through the generations

hoax An act or object passed off as real and meant to fool someone

humanoid Something that is not human, but has the appearance or behavior of one

instinct Something that an animal or a human does naturally, without having to think about it

livestock Animals bred and raised in captivity, usually on farms, that are used for food or some other product, such as milk

omen Something that happens that is believed to be a sign of a future event

plains A large area of flat, treeless land

prey An animal that is killed and eaten by another animal

settler A person who moves to a new place. Also called a colonist.

shapeshifting The ability to take a different form or shape

supernatural A force that is beyond normal understanding

INDEX